California's Complex
Water System

Pamela Brunskill, Ed.M.

Consultants

Kristina Jovin, M.A.T.
Alvord Unified School District
Teacher of the Year

Heather Almer, M.S.
Watershed/Education Specialist

Jessica Buckle
Fullerton School District

Publishing Credits

Rachelle Cracchiolo, M.S.Ed., *Publisher*
Conni Medina, M.A.Ed., *Managing Editor*
Emily R. Smith, M.A.Ed., *Series Developer*
June Kikuchi, *Content Director*
Marc Pioch, M.A.Ed., and Susan Daddis, M.A.Ed., *Editors*
Courtney Roberson, *Senior Graphic Designer*

Image Credits: p.5 William Croyle, California Department of Water Resources; p.6 Library of Congress [LC-DIG-pga-07709]; p.7 Library of Congress [LC-USZC4-11554]; pp.8–9 Photo by G. Thomas; pp.9 (left and right), 31 Security Pacific National Bank Collection/Los Angeles Public Library; p.10 [Two men examining kit of dynamite and wire found during sabotage incidents of Owens Valley Aqueduct, Calif., circa 1924], Los Angeles Times Photographic Archives (Collection 1429). UCLA Library Special Collections, Charles E. Young Research Library, UCLA.; pp.11, 32 Historical Photo Collection of the Department of Water and Power, City of Los Angeles; p.13 (top and bottom) PRISM Climate Group, Oregon State University; pp.14–15 Sierra Club Bulletin, Vol. VI. No. 4, January, 1908, pg. 211; p.16 (foreground) Library of Congress [LC-DIG-ggbain-06861]; pp.18–19 trekshots/Alamy Stock Photo; p.19 (top) Aerial Archives/Alamy Stock Photo; p.20 Jonathan Alcorn/REUTERS/Newscom; pp.22–23 Reed Kaestner/Getty Images; p.26 (bottom) Leigh Green/Alamy Stock Photo; p.28 Bob Kreisel/Alamy Stock Photo; all other images from iStock and/or Shutterstock.

Library of Congress Cataloging-in-Publication Data
Names: Brunskill, Pamela, author.
Title: California's complex water system / Pamela Brunskill, M.Ed. ; consultant, Heather Almer, M.S., watershed/education specialist.
Description: Huntington Beach, CA : Teacher Created Materials, [2018] |
Includes index.
Identifiers: LCCN 2017014112 (print) | LCCN 2017022817 (ebook) | ISBN
9781425854973 (eBook) | ISBN 9781425832452 (pbk.)
Subjects: LCSH: Waterworks--California--Juvenile literature. | Water-supply--California--Juvenile literature.
Classification: LCC TD224.C3 (ebook) | LCC TD224.C3 B78 2018 (print) | DDC
333.91009794--dc23
LC record available at https://lccn.loc.gov/2017014112

Teacher Created Materials
5301 Oceanus Drive
Huntington Beach, CA 92649-1030
http://www.tcmpub.com

ISBN 978-1-4258-3245-2

Printed in China
Nordica.042019.CA21900332

Table of Contents

California's Million-Dollar Water

Water is a precious **resource**. It runs from taps. It fills pools.

Californians need water for everything. Farmers need water for their crops. Businesses need water to make products. People need water for cooking and cleaning. Water allows California to grow and **prosper**. California's **economy** is the largest in the United States. In fact, it is one of the largest in the world, and it's all made possible because of water.

Growing Food

About half of the United States' fruits, vegetables, and nuts are grown in California. That's a lot of food. Farming requires a great deal of water.

California's long coast borders the Pacific Ocean.

But water is not always available. At times, it has to be **conserved**. Some places in the state can get little to no rain for months at a time. Canals and pumps store and move water to cities, farms, and businesses.

Who are the water users? Where does the state's water come from? How is water conserved?

Not Just a Splash!

Oroville Dam is the tallest dam in the United States. At 770 feet (235 meters), it is even taller than the Golden Gate Bridge! The dam controls more than 1 trillion gallons (3.7 trillion liters) of water.

In 2017, heavy rainfall damaged one of Oroville Dam's emergency **spillways**.

Water from Owens Valley!

Through the eighteenth and early nineteenth **centuries**, Los Angeles used its own river for water. As the city grew, that river could no longer support the amount of people who lived there.

Growth of Los Angeles

In 1842, gold was found near Los Angeles. People came in search of wealth. In 1850, the city was home to 1,610 people. Ten years later, 4,385 people lived there. More and more people moved to the city. By the turn of the century, more than 100,000 people had moved to the area!

Businesses boomed. People liked living in the area. But there wasn't a lot of local water. Local streams and wells were being used up. The Los Angeles River was not enough for the growing **population**. The city feared a water crisis.

The people needed more water. Leaders looked for a way to bring water in from outside the city. That way, the city could grow. People could come to Los Angeles and dream of riches.

Fool's Gold

In 1848, the first pieces of gold were found on John Sutter's property. He tried to keep it a secret, but word got out. His land was overrun with men seeking fortunes. He lost everything by 1852, while other people got rich.

Gold!

The discovery of gold near Los Angeles was not huge news. Six years later, in 1848, more gold was found in Northern California. This discovery led to what became known as the Gold Rush.

Los Angeles in the late 1880s

William Mulholland was in charge of getting water for Los Angeles. He was concerned. So was Fred Eaton, the city's **former** mayor. The two men came up with a plan to get water from Owens Valley. This valley is in the eastern and central part of the state. They thought they could bring the water south with an **aqueduct**.

But there was a problem with their idea. People in Owens Valley already had plans for their water. Farmers and ranchers wanted the water for their crops and animals. They wanted to improve their **irrigation** systems.

Shady Tactics

Eaton spoke with people in power. He advised them that the water would be better used in the south. Eaton also bought land and water rights in Owens Valley. The people from Owens Valley thought Eaton was buying those rights for their project. But he wasn't. The rights were for Los Angeles.

The Reclamation Act

In 1902, the Reclamation Act was approved. The law formed a government agency. This group's goal was to help private farmers meet their water needs. But it did not help the people in Owens Valley.

Fred Eaton

William Mulholland

Far-Reaching Friendship

Eaton was Mulholland's boss when they both worked for a water company in Los Angeles. Mulholland began as a ditch cleaner. Mulholland impressed Eaton with his **work ethic**. In turn, Eaton helped him move up in the company.

The Los Angeles Aqueduct

In 1905, Mulholland and Eaton got their way. Plans for an irrigation system in Owens Valley were dropped. And an aqueduct was approved.

In 1908, work began on the aqueduct. It would bring the water 200 miles (322 kilometers) south. The water would come to Los Angeles.

Water flowed all the way to the San Fernando Valley. The Los Angeles Aqueduct was built to supply water for millions of people. It allowed for huge growth.

Explosive Reaction

Back in Owens Valley, people were mad over the sale of their water. Things got worse when their farms were drained. Their land used to be lush and green. But it soon became dry and brown. They showed how mad they were. In 1924 and 1927, **protestors** blew up parts of the aqueduct.

These men study dynamite found after an explosion at the aqueduct.

"Take it!"

In 1913, the L.A. Aqueduct was the largest aqueduct in the world. During the dedication ceremony, Mulholland pointed to water flowing out of it. He told the crowd, "There it is. Take it!"

A group of men help build the aqueduct.

The Water Wars

California's water system uses pipes, tunnels, canals, dams, and pumping plants. It spans remote areas of the state. The system brings jobs to these areas. With such a large setup, the system affects many people. It creates many viewpoints. Owens Valley was not the only time people argued over water rights. It was just the start.

North vs. South

The water wars are usually described as Northern versus Southern California. Most of the water in the state is located in the north. But people in the south use most of the water. People who live in the north are concerned. They fear the state favors bringing their water to people and businesses in the south. People in the south claim that they need water to survive and for the state to thrive.

grapevines in Northern California

lettuce in Southern California

California Crops

People in the United States rely on California's crops. The state is the leading producer of grapes, lettuce, avocados, tomatoes, and strawberries. Nuts are big business for many farmers. Almost all of the country's almonds, pistachios, and walnuts are grown in California.

Economics

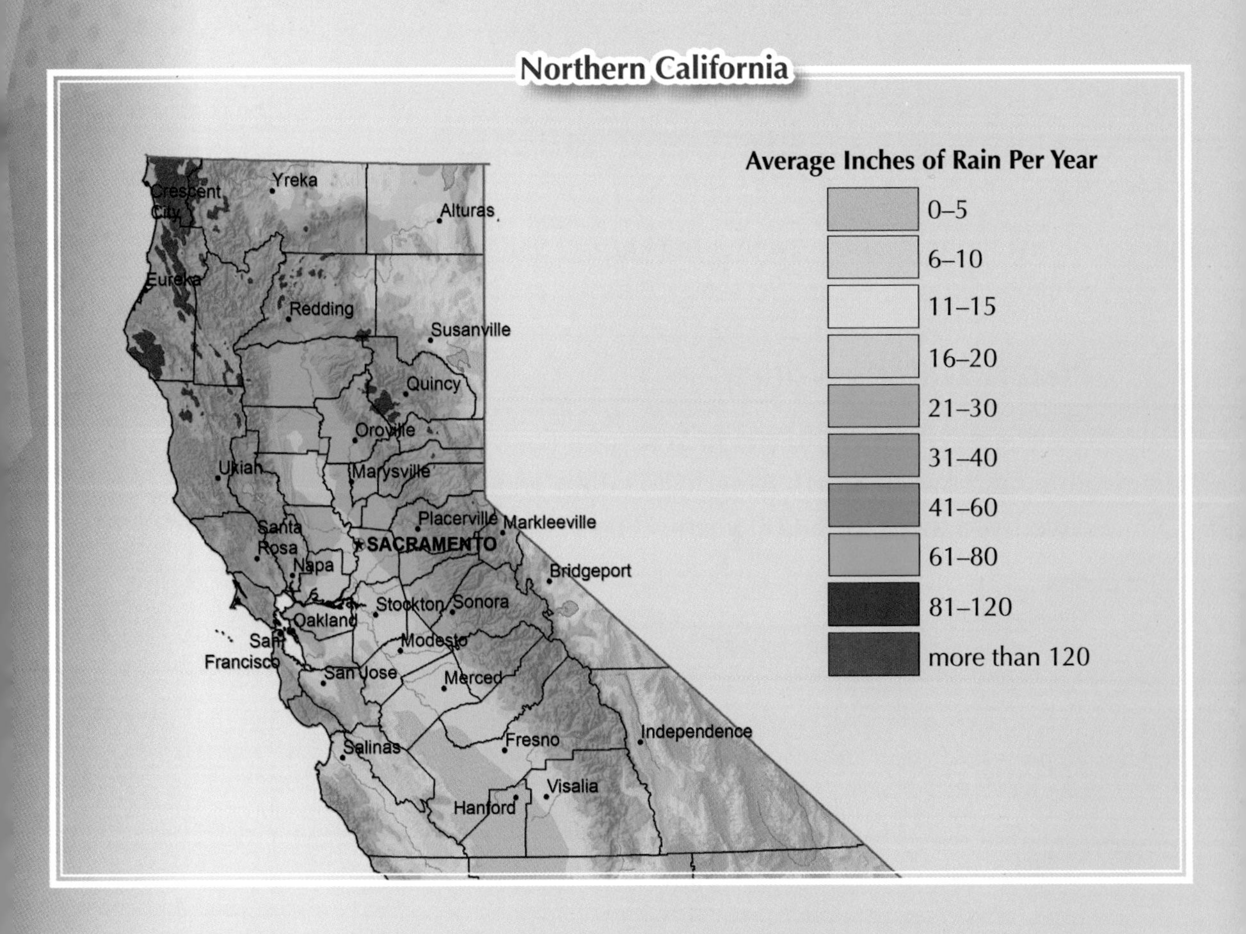
Northern California
Average Inches of Rain Per Year
0–5
6–10
11–15
16–20
21–30
31–40
41–60
61–80
81–120
more than 120
Crescent City
Yreka
Alturas
Eureka
Redding
Susanville
Quincy
Oroville
Ukiah
Marysville
Placerville
Markleeville
Santa Rosa
SACRAMENTO
Napa
Bridgeport
Stockton
Sonora
Oakland
San Francisco
Modesto
San Jose
Merced
Independence
Salinas
Fresno
Visalia
Hanford

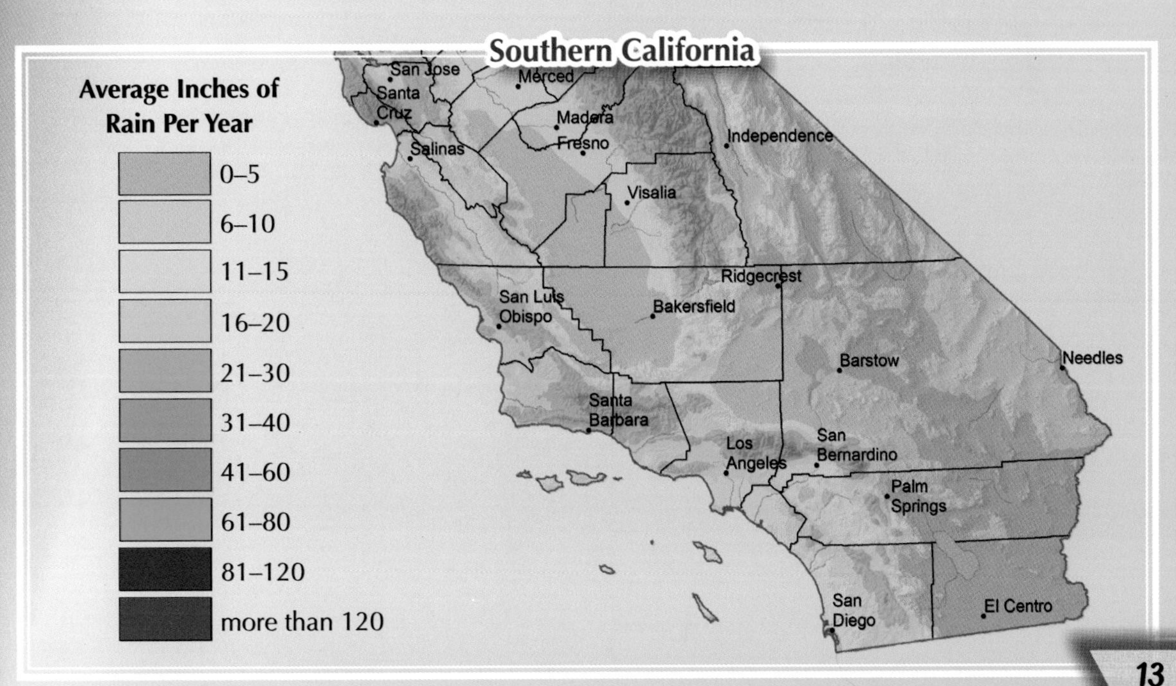
Southern California
Average Inches of Rain Per Year
0–5
6–10
11–15
16–20
21–30
31–40
41–60
61–80
81–120
more than 120
San Jose
Merced
Santa Cruz
Madera
Fresno
Independence
Salinas
Visalia
Ridgecrest
San Luis Obispo
Bakersfield
Barstow
Needles
Santa Barbara
Los Angeles
San Bernardino
Palm Springs
San Diego
El Centro

Each region has a point. The fight is about where the water is and where it is needed. And water wars are not just between the north and the south.

At the turn of the twentieth century, San Francisco was growing fast. It needed more water. So the city looked east.

Hetch Hetchy Valley

Hetch Hetchy Valley is in Yosemite National Park. It has **granite** cliffs and two of North America's tallest waterfalls. American Indians have lived there for 6,000 years. The valley is a natural gem.

This photograph shows the beauty of the valley before the dam was built.

Some people wanted to build a dam in Hetch Hetchy to create a **reservoir** for the city. How much would it cost to get the water? In this case, cost was not about money. There was a huge cost to the environment. Building a dam would devastate the valley. Cutting down all the trees and flooding the valley would destroy the plants and animals.

Pricey Water

After the Gold Rush, San Francisco had to buy water to meet the needs of its people. At the time, it cost $1 a bucket. That is about $26 per bucket today!

Economics

Another Loss

American Indians were concerned with the impact the dam would have on their lives. They would lose their hunting grounds. Many of the plants they ate and used to make baskets would be gone. The native people's sacred burial grounds would also be disturbed.

Geography

To Conserve or Preserve

The U.S. government protected Hetch Hetchy Valley. But people still fought over the land. **Conservationists** said the valley should be used to help people. San Francisco residents argued for a dam. They said it would help keep their city strong.

Sierra Club

John Muir founded the Sierra Club in 1892. With this group, he urged people to write to elected officials to oppose the dam in Hetch Hetchy. The Sierra Club lost the battle.

Preservationists said the land should be protected. John Muir led this group. He thought Hetch Hetchy should be enjoyed for its beauty alone. Both sides wrote to Congress for help.

In 1913, Congress ruled that the dam could be built. Hetch Hetchy Valley would supply San Francisco and the Bay Area with water. The region's potential for growth won out over the environment. The project took 25 years to finish. San Francisco still receives water from the dam.

Hetch Hetchy Reservoir

National Park System

The efforts to save Hetch Hetchy Valley made people start to think about preserving parks. It helped create the National Park Service. This group was formed in 1916.

Civics

The Water System and the Economy

California has a big water system. The system includes many aqueducts. It is one of the biggest in the country. Lots of money is involved in running it. So, it has a big impact on the state's economy.

The Hetch Hetchy dam and reservoir cost over $100 million to build. It brought jobs to nearby areas. Just like the Los Angeles Aqueduct, it allowed for growth. With the system in place, more people and businesses could come to San Francisco.

The same is true with all water systems. People are needed to build them, run them, and repair them. When systems are in place, businesses can grow. This is key in cities where water is scarce.

Sides of the Water Wars

The water wars are not just between the North and the South. They are also between the people who want to conserve and those who want to preserve. Further, they are between **rural** areas versus cities.

Pump It Up!

Getting water requires a lot of **energy**. Water needs to be pumped, treated, and supplied to people. About 20 percent of the state's energy is focused on getting water.

water treatment plant

The Cost of Water

It is expensive to move water from one place to another. It costs the state $600 million each year to run the water system. The system brings water to more than half the people in the state.

Economics

Throughout the state, aqueducts, such as this one, flow from the mountains to the cities.

Water Systems and the Environment

California's water system has helped the state to grow and prosper. This is a good thing. But building dams, pipes, and water treatment plants affects the environment. This is not always good.

In 1941, the Los Angeles Aqueduct stretched north of Owens Valley. It went to Mono Lake. Over the next 40 years, the lake's water level dropped. It lost half its volume. It doubled its salt levels. This hurt the **ecosystem**. Ducks and geese stopped coming to the lake. The fish population declined.

Today, people concerned about the environment fight against new dams. They say dams have caused enough damage. State leaders are trying to balance the needs. They must consider the needs of humans, the economy, and the ecosystem.

Power welders repair a pipe that burst.

Maintain the System

Many people rely on California's water system. Water mains and lines require repairs or upgrades. With routine maintenance, people in the state will continue to have clean, fresh water.

Saltwater Lake

Mono Lake is a saltwater lake near Yosemite National Park. It's known for the many unique rock formations that surround the lake. Water levels at the lake dropped to dangerous levels because of the **drought**. But heavy rains in 2016 and 2017 have helped. The lake's water level rose significantly!

At Mono Lake, the plants are covered in salt.

Alternative Solutions

Experts are looking for new ways to bring water where it is needed. Taking salt out of salt water and recycling water are two options.

Drink the Sea?

California lies next to the Pacific Ocean. But ocean water is not drinkable. So, scientists have been working on turning salt water into fresh water.

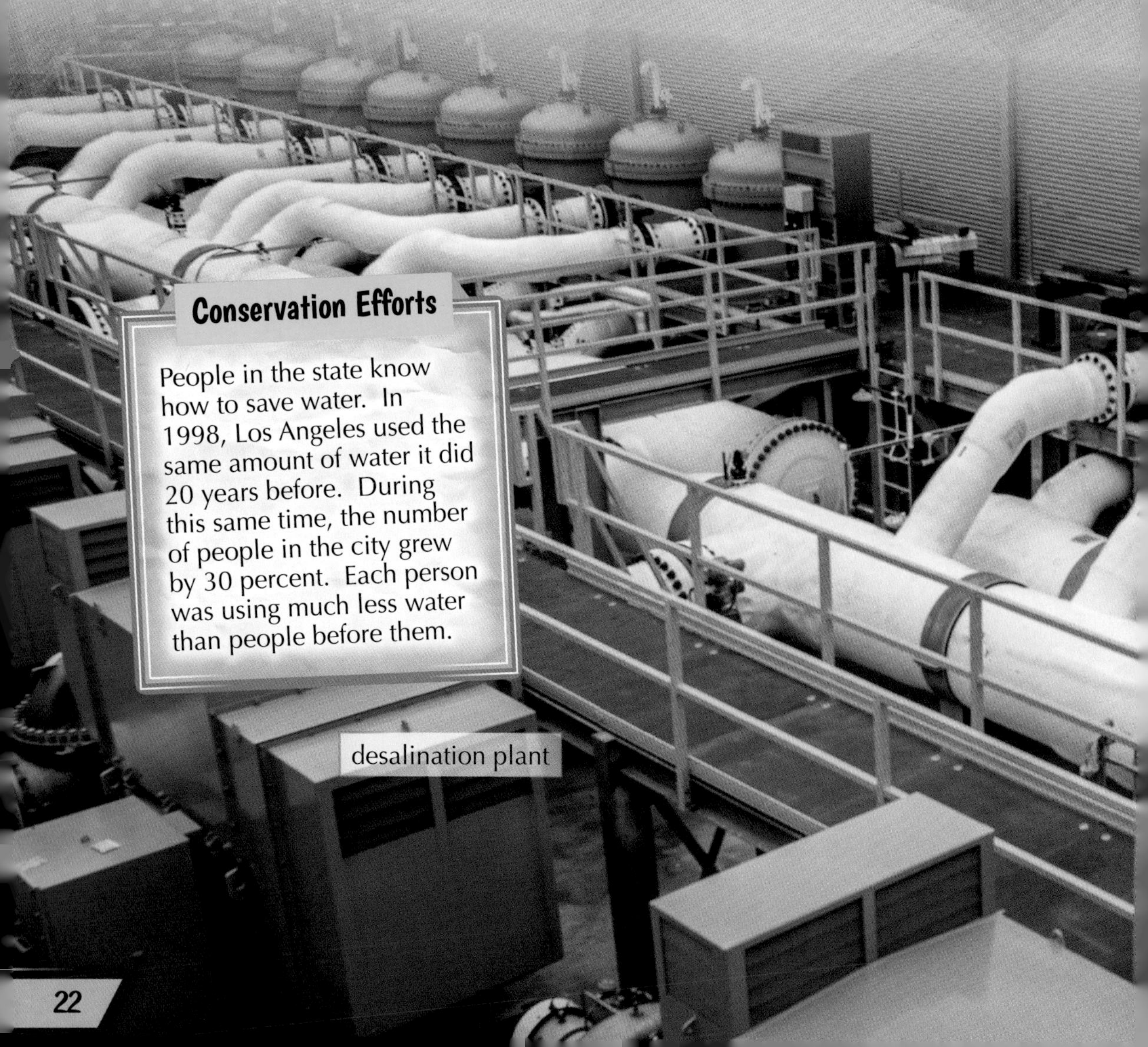

Conservation Efforts

People in the state know how to save water. In 1998, Los Angeles used the same amount of water it did 20 years before. During this same time, the number of people in the city grew by 30 percent. Each person was using much less water than people before them.

desalination plant

Desalination is the process of removing salt from water. It is not new. Since ancient times, people have been doing this. The most basic way to do this is to boil salt water in a pan, capture the steam, and allow that steam to turn back into water. That water is drinkable.

Today, taking salt out of ocean water on a large scale is expensive. It uses a lot of energy and involves large facilities. But it can be done. San Diego County built a desalination plant in 2015. It cost $1 billion. The plant helps shield the city from drought.

Capturing Storm Water

Another way the state is trying to get water is by capturing storm water. In typical drains, rain goes down and into the ocean. In special drains, water goes into the soil and groundwater.

a modern tank used to collect rainwater

Recycled Water

Do you want to drink water from sewers? How about from urine? That may sound gross, but it can be done.

California uses several steps to clean this kind of water. It is called water treatment. Wastewater flows through screens. This removes large objects. Then, it flows through tanks. There, heavy things in the water sink and light things float. Then, the water gets **disinfected**. This kills bad **bacteria**. After water goes through this kind of treatment, it is drinkable.

A Fair Price

Leaders in California must think more about how much water is needed and how to get it. People need water. Land and animals need water. Businesses need water. There is no easy answer to the state's water issue.

The number of people in the state is still growing. Given the needs of humans and the environment, what is a fair price for water?

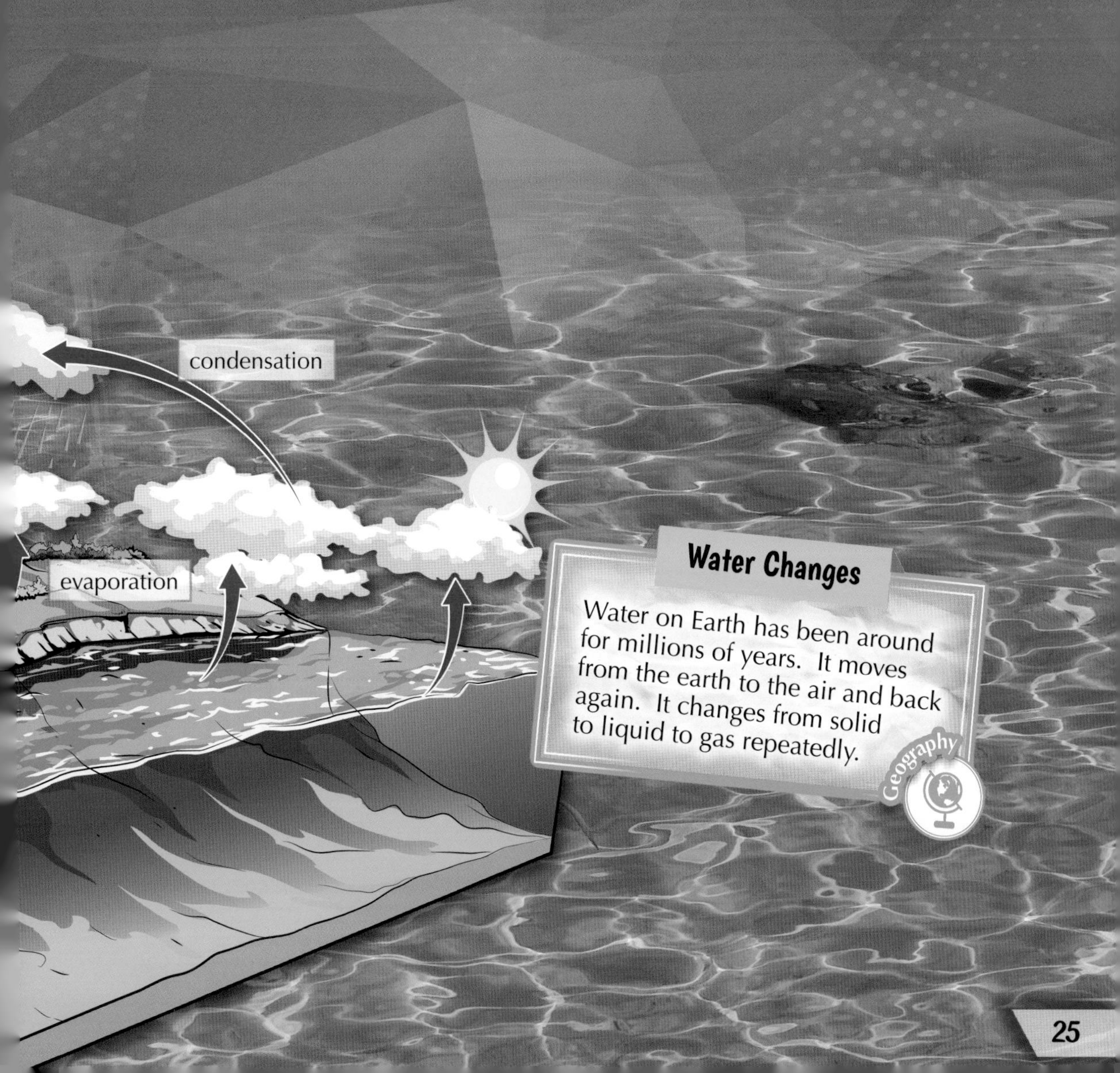

Water Changes

Water on Earth has been around for millions of years. It moves from the earth to the air and back again. It changes from solid to liquid to gas repeatedly.

Geography

California's Water Future

Californians have been conserving water for decades. They use low-flow toilets. They buy machines that don't use a lot of water. They follow laws for how to use water. These efforts are not enough.

The state's aqueducts are not being filled as fast as they are being used. And dry conditions have plagued the state over the years. In 2016, California entered its fifth straight year of drought. The state has endured droughts in the past. This trend will likely continue in the future.

The drought affects people and businesses. Agriculture, energy, and tourism are just a few industries that are hurt by drought. The Golden State must do more to meet its need for water.

Rules

People in California are used to drought conditions. Because of this, they use water wisely. But, there are still regulations for water usage. Each city has its own rules.

Water Limits

In 2015, Governor Jerry Brown told people in the state to reduce their water use. The state's government website has news and information about water. Go to drought.ca.gov and check it out.

Lake Mead is a reservoir from which California draws water. It was at very low levels during the drought.

Debate It!

What are some solutions to the water problem in California? Choose one possible solution that you think would help meet the needs of the state's people, businesses, and environment.

Develop an argument about why California should invest in your idea. Use what you have learned about the history of the water wars to make your case stronger.

Make a list of the pros and cons of your solution.

A sign shows that farmers and their crops are severely affected by droughts.

lows®
alition
District

Glossary

aqueduct—a human-made canal that brings water from one place to another

bacteria—small living things that can cause diseases

centuries—periods of 100 years

conservationists—people who work to protect natural resources for later use

conserved—used carefully and not wasted

disinfected—cleaned; destroyed germs

drought—a long period of dry weather

economy—the system of buying and selling goods and services

ecosystem—living and nonliving things in a certain environment

energy—power that comes from a source, such as heat, electricity, wind, or the sun

former—describes what someone or something was in the past

granite—a very hard rock that is used in monuments, buildings, and homes

irrigation—the act of supplying lands with water by taking it from streams, lakes, or rivers

population—the total number of people or animals in an area

preservationists—people who work to keep lands safe and in good condition

prosper—become successful, usually by making money

protestors—people who say or do things that show disagreement with something

reservoir—a lake used to store a large supply of water for use

resource—something that a country has that can be used

rural—relating to the countryside or area outside a town or city

spillways—structures built to provide a safe release of flood waters

work ethic—a strong belief in the value of work

Index

Your Turn!

Help Wanted

People in Owens Valley wanted the water to stay in the valley. People in Los Angeles argued that they needed the water.

Imagine you were living at this time. Choose one side to support. Create a poster that explains why it's important for your side to get the water from Owens Valley. Make your poster attractive by using visuals.